An Introduction to SI Units

for students of science and technology

An Introduction to SI Units

for students of science and technology

J. A. Glenn

Kesteven College of Education

Heinemann Educational Books Ltd
London

Heinemann Educational Books Ltd
LONDON EDINBURGH MELBOURNE TORONTO
AUCKLAND SINGAPORE JOHANNESBURG
HONG KONG IBADAN NAIROBI NEW DELHI

ISBN 0 435 68170 2

Published by Heinemann Educational Books Ltd
48 Charles Street London W1X 8AH
Printed in Great Britain
by Spottiswoode, Ballantyne & Co. Ltd.
London and Colchester

Preface

THIS is, by intention, a 'book of the hour'. It is offered solely to meet the needs of one class likely to be inconvenienced by the transition to SI units in science and technology. Students and young people in school, faced with examinations set entirely in the internationally agreed system, may find difficulty if they have already become familiar with working in the units which are shortly to be regarded as obsolete. Whatever the practical and organizational difficulties in industry and commerce, here at least is a clearly defined problem. This booklet is an attempt to solve it.

There is no question of presenting a case. The arguments have been marshalled and the decisions have been taken: our task in the schools and colleges is to put the decisions into practice as quickly and effectively as possible.

I am convinced that after the transition has taken place the various mathematical and indeed conceptional difficulties that arise because of our present multiplicity of units will vanish, along with the chapters on units so often found in textbooks. I look forward to the time when each unit is introduced as the developing subject matter requires it, when it will be used as needed but never appear as a wholly artificial topic in the arithmetic or physics of conversion, such as arises if we express horsepower in watts or farads as centimetres.

I am also convinced that, even for the student who is half way through a course, the new look at units should be taken in a direction entirely forward. One should not recapitulate the familiar units and show how they can be converted to the new; but one should present the SI units simply and directly and show how they enable one to handle quantitatively the phenomena which are studied in science and put to practical use in technology. This is what I have tried to do.

Given an adequate knowledge of his subject matter, the student who has read this work attentively should be able to solve, in SI units, numerical problems whose data are in these units. I would suspect that any difficulties that a few practice examples did not dispel arose from a more serious confusion than in the units of measurement. The whole point of the SI units is their simplicity and consistency. Properly understood, they remove a barrier between various branches of science and technology: the language of measurement becomes a common language.

For the reader who needs more information about the system than suffices for the limited but, I believe, important objective of this booklet, a short bibliography is added. For the sixth former or student in a college of further education the work should be complete in itself. The examples are simple and direct calculations taken from various branches of science or industry merely to show the units in use. They are *not* intended to illustrate details of the subject matter, although it is hoped that they are set out sufficiently fully to be complete without reference to text-books. A number of unworked examples is added for further practice. I hope that students will attempt a few taken from fields other than those in which they are studying, and in this way convince themselves of the value of a uniform system.

I have taken the opportunity offered by a second printing to remove a number of mistakes pointed out to me by readers, and to revise several passages concerning units etc., on which no authoritative decision had been made at the time of first publication.

My thanks are due to my colleague Mr. K. Lewis and to Messrs. S. S. M. Nussey and S. J. Branson, both of Newton's old school at Grantham, for help in checking the manuscript. They are in no way responsible for residual errors.

J.A.G.

Contents

Preface v
An international system of units 1
Space, time and motion. 4
Mass, force and energy 9
Heat and thermodynamics 14
Illumination 20
Electrical phenomena 22
Appendix A—A note on dimensions 34
Appendix B—Measurement in atomic physics 37
Appendix C—Measurement in acoustics 39
Appendix D—Miscellaneous examples 42
Bibliography 49
Answers 51
Index 52

1

An International System of Units

We can accept that units of measurement now in use represent the end product of a historical process rather than an attempt to construct a system adapted to current or foreseeable needs. In particular, older units tend to be restricted in use to limited fields—to science but not to technology, to England but not to Germany and so on. There has long been the need for a standard international system of measures.

This has now been provided by the Système International d'Unités (or SI), adopted as the sole legal standard by many countries using the former metric system, and now being adopted, pending legislation, by industry and science in most others. It is already, in effect, the system used in industrial electrotechnology. It will be used in most examinations for certificates in general or secondary education before or by 1972, and will eventually replace other units at all levels of instruction.

The system does not construct new units: it merely selects, redefines and develops a set of measures adequate for all purposes in science or technology. Its novelty consists only in its careful attention to details of its construction. In the past units have, like the horsepower, been introduced *ad hoc*. The design of SI is covered by five simple principles:

1. The number of fundamental, independent units is kept to the minimum consistent with convenience.

2. All other measures, of force, speed, density and so on are *derived units*, constructed from the fundamental units without numerical factors or arbitrary constants. The system is said to be *coherent*.

3. Derived units in common use are given internationally agreed names for convenience. (Most named units are already familiar.)

4. The multiples and submultiples of units required to give them an adequate range are restricted by recommendation.

5. Certain units already in international use are retained as 'customary units' but do not form part of the system.

The pages that follow describe the SI units *in use*. They do not discuss the formal definitions beyond some indication of their nature. Basically, the SI system is a metric system, but is best learnt by using it without reference back to its historical origins.

Common to all the units of the system are the prefixes indicating multiples and submultiples, adapted from those used in the original metric measures. Those recommended represent powers of ten where the index is itself a multiple of three, although two others may appear as follows:

Prefix	Value	Symbol
giga-	$\times 10^9$	G
mega-	$\times 10^6$	M
kilo-	$\times 10^3$	k
(deci-)	$\times 10^{-1}$	(d)
(centi-)	$\times 10^{-2}$	(c)
milli-	$\times 10^{-3}$	m
micro-	$\times 10^{-6}$	μ
nano-	$\times 10^{-9}$	n
pico-	$\times 10^{-12}$	p

Of these deci- and centi- (in parentheses) are not recommended for general use, but may as a matter of custom and convenience appear in certain contexts. These and other customary usages will be discussed as they arise.

All units are represented by international symbols, in roman type. These are not abbreviations and do not therefore take a full stop (except at the end of a sentence) or any inflexion for the plural.

The symbols and notation used in this booklet follow BS 1991 and the publications of the Association for Science Education, the latter being based on the recommendations of the Royal Society. The only significant difference between these two sets of recommendations is in the use of negative indices rather than the solidus when writing derived units, e.g. $\mathrm{m\,s^{-1}}$ instead of m/s.

The Royal Society follows growing international scientific practice in using negative indices, while BS 1991 follows the firmly established custom in industry and technology and employs the solidus. It seems clear that both notations will persist, particularly as there is no difference in principle between them. The solidus is certainly easier to write and type, but it still needs positive indices in expressions such as acceleration in metre per second squared, written in $\mathrm{m/s^2}$ in the B.S. notation and $\mathrm{m\,s^{-2}}$ in the Royal Society usage. Students should accustom themselves to both forms.

This booklet follows the index form (which is likely to be adopted in schools) mainly because it is at the moment the least familiar to students. The solidus is, however, likely to remain general in customary usage in such expressions as kilometre per hour, km/h, and revolution per minute, rev/min, and is so used in this text.

Where units are written in full plural inflexions are omitted from the names as from the symbols.

In Chapter 6 I have, since I am merely concerned with the use of units, discussed all four of the quantities D, E, B, and H, although school courses may adopt a different approach. All four are commonly used in technical courses.

2

Space, Time and Motion

ALTHOUGH astronomers often use the speed of light as a unit and hence derive the 'light year' as a unit of distance, it is generally more convenient to have units of length and time as starting points.

SI defines a single universal unit of length, the *metre*, symbol m, and a single unit of time, the *second*, symbol s. The metre is defined in terms of the wavelength of the light from a prominent spectral line in an isotope of the gas krypton; the second astronomically as a fraction of a particular tropical year, standardized from now on against the frequency of a convenient spectral line of caesium.

The recommended multiples for the metre are

kilometre	km
millimetre	mm
micrometre	μm

The term *micron*, μ, for micrometre will, it is hoped, disappear. The centimetre, cm, is a permissible usage. It is better to use a recommended unit and write 157 mm rather than 15·7 cm in recording and so avoid possible mistakes with the decimal point.

Although the second may be used with the standard prefixes the system, sensibly adapting itself to ordinary usage, allows the use of minutes, hours, days, years where these are obviously called for. This usage, however, introduces such factors as 60, 3600, 24 into calculations which pure SI would avoid.

The recommended multiples and submultiples will be used for practical measurement, calibration, and technological specification, but for purposes of computation in science it is usually more convenient to use the metre and the second only, writing results with powers of ten. This can be either the usual 'standard form' where a quantity is given as $a \times 10^n$ where $10 > a > 1$ and n is a positive or negative integer. Alternatively n can be chosen to make a an integer and hence not require a decimal point. Convenience is a better guide than consistency. $27\cdot5\,\text{km}$ is better than $275 \times 10^{-1}\,\text{km}$.

Here are some examples of typical measures in length and time using permissible SI units, illustrating variations that may be found in practice.

One light-year	$9\cdot46 \times 10^{12}\,\text{km}$
	$9\cdot46 \times 10^{15}\,\text{m}$
Distance of the moon	$384\,000\,\text{km}$
Radius of earth	$6378\,\text{km}$
Height of a man	$1750\,\text{mm}$
Thickness of cigarette paper	$25\,\mu\text{m}$
Effective diameter of water molecule	$4\cdot1 \times 10^{-10}\,\text{m}$
Wave length of sodium lines D_1	$5896 \times 10^{-10}\,\text{m}$
D_2	$5890 \times 10^{-10}\,\text{m}$
Half-life of radium (from uranium)	1590 years
Half-life of polonium (from thorium)	$3 \times 10^{-7}\,\text{s}$

Our decimal point is not an international notation. Continental countries tend to use the comma, e.g. $4,21$ not $4\cdot21$. It is now general practice to use spacing and not commas in numbers of more than 4 digits, e.g. $52\,400\,000$ not $52,400,000$, although one writes 4 digit numerals without space or comma.

Where magnitudes differ considerably it is better to use the metre only for computation.

Example 1

Compare the thickness of a cigarette paper with the diameter of a molecule of water.

Our table gives one in micrometre, the other in metre. We use the metre.

Then

$$\text{Thickness of paper} = 2\cdot5 \times 10^{-5}\,\text{m}$$

$$\text{Diameter of molecule} = 4\cdot1 \times 10^{-10}\,\text{m}$$

$$\text{Hence ratio} = \frac{2\cdot5}{4\cdot1} \times 10^5$$

$$= 6\cdot1 \times 10^4$$

$$= 61\,000$$

The effective diameter of a water molecule is less than $1/60\,000$ part of the thickness of a cigarette paper.

Velocity and *acceleration* now appear as derived units of the system, measured respectively in

metre per second	m s^{-1}
metre per second squared	m s^{-2}

The notation m/s, m/s^2 will also be found.

Kilometre per hour (km/h) passes into SI without comment, but cm/s is a permitted rather than recommended usage. This point need not be laboured.

Example 2

A car attains a speed of 40 km/h from rest in 7 s. What is its average acceleration?

$$\text{Final velocity} \quad = \frac{40\,000}{3600}\,\mathrm{m\,s^{-1}}$$

$$\text{Hence acceleration} = \frac{400}{36 \times 7}\,\mathrm{m\,s^{-2}}$$

$$= 1{\cdot}59\,\mathrm{m\,s^{-2}}$$

To handle problems in curvilinear motion, rotation, rolling and so on SI needs a measure of angle. The sexagesimal degree (1 complete turn = 360°) is internationally known and appears as a customary unit, but is not taken as fundamental. This is partly due to the existence of an alternative centesimal system (1 complete turn = 400 grades; 1 grade = 100 centigrades) often used on the continent, but mainly to the greater convenience in theoretical calculation of an angular measure derived directly from length. SI takes the *radian* (rad) as the angle subtended at the centre of a circle by an arc equal in length to the radius. The unit is thus a numerical ratio, metre/metre. Since one complete turn takes a point on the circumference completely round a distance $2\pi \times$ radius

$$2\pi\,\mathrm{rad} = \text{one complete turn}$$

Example 3

If a road wheel of the car in Example 2 has a diameter of 570 mm, find the average angular acceleration of the wheel.

$$\text{Circumference of wheel} = (\pi \times 0.57)\,\mathrm{m}$$

$$\text{Hence at 40 km/h its rotation} = \frac{40\,000}{3600 \times \pi \times 0.57}\,\mathrm{rev/s}$$

$$= \frac{400 \times 2\pi}{36 \times \pi \times 0{\cdot}57}\,\mathrm{rad\,s^{-1}}$$

$$= 39\,\mathrm{rad\,s^{-1}}$$

Hence average angular acceleration from rest

$$= \tfrac{39}{7}\,\mathrm{rad\,s^{-2}}$$

$$= 5{\cdot}57\,\mathbf{rad\ s^{-2}}$$

The radian is regarded as a supplementary unit of the system.

It is likely that SI will eventually recognize another customary unit of angle. This is the *mil* which is 1/6400 of a complete turn. It is now used generally by the armed forces for gun laying.

It is worth noting that for the calculations required by astronavigation the units at present in use, the degree, the nautical mile and the knot, will *not* be replaceable by SI units. This is because there is a simple arithmetical relationship between time and the earth's angular velocity (it rotates 15 degrees in 1 hour) and between angle and distance on the earth's surface (1 degree difference of latitude along a meridian $\equiv$ 60 nautical miles). Appendix B discusses another special set of units.

Measures of surface area and volume use the *square metre* (m^2) and the *cubic metre* (m^3). The regular multiples and submultiples of these are not always convenient for recording purposes, and the 'customary' units cm^2, cm^3, and dm^3 are likely to continue in use. For liquid measure the *litre* (l) and the millilitre (ml) will continue in use, but do not appear as a formal part of the system. There is a sound historical and practical reason for this exclusion (see bibliography). Land surveying, for which the square kilometre is too big and the square metre too small, will use the *hectare* (ha) a measure taken direct from the former metric system and defined as $10\ 000\,m^2$.

Example 4

Calculate the total volume of water which falls on a field of area 2·37 ha during a day when the rainfall is 19 mm.

$$\text{Volume of water} = 2·37 \times 19 \times 10\,m^3$$

$$= 2·37 \times 19 \times 10^4 l$$

$$= 450·3\,kl$$

These 'customary' uses are not governed by rules but by convenience. It is not, for example, convenient to use the centimetre in giving dimensions if subdivision is needed, since this would need a decimal point. Thus one would specify a rectangular opening in a structure as $125\,mm \times 266\,mm$ and not $12·5\,cm \times 26·6\,cm$.

A measure often used in theory is that of solid angle. SI specifies, again as a supplementary unit, the *steradian* (sr). This is the solid angle subtended at the centre of a sphere by any portion of its surface equal in area to the square on the radius, and is thus the three dimensional equivalent of the radian. Since the surface area of a sphere is given by $4\pi r^2$ it follows that

$$4\pi\,sr = 1 \text{ complete sphere}$$

Note that, although a cone has a solid angle at its vertex, this is not the same as the 'angle of the cone', which is double the plane angle between a generator and the axis. A solid angle can be subtended by any irregular shape: it is only the area intercepted on the sphere that determines the value.

Example 5

A circular hole of radius 3 cm is cut in a spherical shell of radius $r = 5$ cm. Calculate the solid angle of the cone of light emerging from the hole when a uniformly emitting point source is placed at the centre of the shell.

From the dimensions given the plane of the hole is 4 cm from the centre of the sphere and hence the depth of the cap cut off is 1 cm. Its area is, by a familiar theorem, that of its intercept on the cylinder having the same axis of symmetry in which the sphere is inscribed.

$$\text{Hence area of cap} = \frac{4\pi \times 5 \times 5}{10} \text{ cm}^2$$

$$\text{and solid angle of cone} = \frac{\text{area of cap}}{r^2}$$

$$= 4\pi/10 \text{ sr}$$

$$= 1\cdot26 \text{ sr}$$

Frequency, the rate of repetition of a periodic phenomenon such as an oscillation or pulse, is usually measured in cycles per second. SI advocates the use of the name *hertz* (Hz), one hertz being one cycle per second. This is already widely used.

The common multiples are

$$\text{kilohertz} \qquad \text{kHz}$$

$$\text{megahertz} \qquad \text{MHz}$$

3
Mass, Force and Energy

THERE is a large group of words commonly used in science and technology whose exact meanings and interconnections are far from straightforward. Examples from this group are: mass, force, weight, energy, impulse, action, inertia, power, and momentum. In so far as some of these can be defined by the methods used for measuring them, they can be made technically precise although their everyday uses in ordinary speech remain cheerfully inconsistent. The basic intuitive concept is that of force, the exertion of our own muscles, but it is more convenient to establish a unit of *mass*. The SI unit is the *kilogramme* (kg), defined as that of an international prototype kept at Sèvres in France. It is assumed that masses can be compared by using a balance. The larger unit, the megagramme (Mg) is called a *tonne* on the Continent, but this name is not recommended in English while the imperial ton continues in use. The submultiples are:

gramme	g
milligramme	mg
microgramme	μg

Example 6

A sample of ceramic material $40\,\text{mm} \times 25\,\text{mm} \times 15\,\text{mm}$ weighs $32 \cdot 3\,\text{g}$. What is its density?

$$\text{Volume} = 40 \times 25 \times 15\,\text{mm}^3$$

$$= 15\,\text{cm}^3$$

$$\text{hence density} = \frac{32 \cdot 2}{15}\,\text{g\,cm}^{-3}$$

$$= 2 \cdot 15\,\text{g\,cm}^{-3}$$

$$= 2150\,\text{kg\,m}^{-3}$$

Clearly values in $\text{kg\,m}^{-3}$ are more relevant to large scale operations, and it will normally be more convenient to tabulate these. In the example it was the small specimen that was weighed in gramme.

Example 7

A blast wall for a turbine test bed is $20\,\text{m} \times 5\,\text{m} \times 1\,\text{m}$ in reinforced concrete. What is its mass?

Using the tabulated density $2350\,\text{kg}\,\text{m}^{-3}$

$$\text{Mass of wall} = 20 \times 5 \times 1 \times 2350\,\text{kg}$$

$$= 235\,\textbf{Mg}$$

Note that the word 'weight' would be used here in ordinary speech. Our intuitive perception of mass is the *force* our muscles exert in moving or lifting a mass, or arresting a mass in motion. SI gives force a precise measure and definition as a derived unit. A unit force is that which acting on a mass of $1\,\text{kg}$ produces an acceleration of $1\,\text{m}\,\text{s}^{-2}$. This unit is named the *newton* (N). Its recommended multiples are:

$$\text{meganewton} = \text{MN}$$

$$\text{kilonewton} = \text{kN}$$

$$\text{millinewton} = \text{mN}$$

Example 8

The driver of the car in Example 2 has a mass of $76 \cdot 5\,\text{kg}$. What is the average force exerted on him by the seat back during the acceleration?

$$\text{Force} = \text{mass} \times \text{acceleration}$$

$$= 76 \cdot 5 \times 1 \cdot 59\,\text{kg}\,\text{m}\,\text{s}^{-2}$$

$$= 112\textbf{N}$$

A very important natural constant is the force exerted by gravity near the earth's surface on a mass. The measured value, influenced by the eccentricity, the rotation and the local density of the earth, varies from about $9 \cdot 78\,\text{N}\,\text{kg}^{-1}$ at the equator to about $9 \cdot 83\,\text{N}\,\text{kg}^{-1}$ at the poles. From the definition of force, it follows that the acceleration g of a falling body is numerically equal to this. The value of g is usually taken as $9 \cdot 81\,\text{m}\,\text{s}^{-2}$, and its standard value as $9 \cdot 806\,65\,\text{m}\,\text{s}^{-2}$.

It is essential to note that in using SI, which defines force in terms of unit acceleration, the constant g must appear in all problems involving gravitational forces, but in no others.

Example 9

What is the pressure exerted by the blast wall in Example 7 on the concrete raft upon which it rests?

$$\text{Mass of wall} = 235\,\text{Mg}$$

$$\text{Base area} = 20\,\text{m}^2$$

$$\text{Hence pressure} = \frac{235\,000 \times 9 \cdot 81}{20}\,\text{N}\,\text{m}^{-2}$$

$$= 115 \cdot 4\,\text{kN}\,\text{m}^{-2}$$

Care must be taken when both gravitational and non-gravitational forces are involved.

Example 10

A steel wire of unstretched length 3 m and diameter 1 mm supports a mass of 5 kg (i.e. a 5 kg 'weight' is hung on the end). Calculate the extension, given that it is within the elastic limit.

If the extension is e, then

$$\frac{\text{extending force}}{\text{area of cross section}} = \frac{Ee}{\text{length}}$$

Where E is Young's modulus tabulated as $2.1 \times 10^{11}\,\text{N}\,\text{m}^{-2}$.

$$\text{Hence } \frac{5 \times 9.81}{\pi \times 0.5^2 \times 10^{-6}} = \frac{2.1 \times 10^{11} \times e}{3}$$

$$e = \frac{5 \times 9.81 \times 3}{3.14 \times 0.25 \times 2.1 \times 10^5}\,\text{m}$$

$$= 0.9\,\text{mm}$$

This is a more convenient form for the result than $9 \times 10^{-4}\,\text{m}$.

In SI units the pressure of the standard atmosphere is 101.3 kN m^{-2} but barometer readings will doubtless continue to be recorded in millimetre of mercury or in bar

$$1\,\text{mm}\,\text{Hg} = 133.3\,\text{N}\,\text{m}^{-2}$$

$$1\,\text{bar} = 10^5\,\text{N}\,\text{m}^{-2}$$

Given the units of mass, length, time, and force the units for work and power are derived directly. The *joule* (J) measures the work done when the point of application of a force of one newton moves through one metre, and the *watt* (W) measures the power or rate of doing work. One watt is one joule per second.

Note that these derived units can always be written out in full:

$$1\,\text{N} \equiv 1\,\text{kg}\,\text{m}\,\text{s}^{-2}$$

$$1\,\text{J} \equiv 1\,\text{kg}\,\text{m}^2\,\text{s}^{-2}$$

$$1\,\text{W} = 1\,\text{kg}\,\text{m}^2\,\text{s}^{-3}$$

$$\text{and also } 1\,\text{J} \equiv 1\,\text{W}\,\text{s}$$

For most applications the prefixes mega-, kilo-, and milli- will suffice.

Example 11

What is the effective work done if 15 kl of water is pumped through a height of 20 m?

The work is done against gravity and hence, since the mass of 1 l of water is 1 kg for all practical purposes,

$$\text{work done} = 15 \times 10^3 \times 9\cdot81 \times 20 \; \text{kg m}^2\text{s}^{-2}$$

$$= 2\cdot94\,\mathbf{MJ}$$

Example 12

A train travels at 40 km/h and the drawbar pull at this speed on a level track is 22·3 kN. What power is being developed by the locomotive?

$$\text{Power} = \frac{22\cdot3 \times 10^3 \times 40 \times 10^3}{3600} \; \text{W}$$

$$= 248\,\mathbf{kW}$$

Example 13

A gear wheel of diameter 1500 mm is driven at 150 rev/min by another gear wheel providing a force of 9·1 kN at the pitch line. Find the torsional moment on the shaft and the power transmitted.

$$\text{Moment} = \text{radius} \times \text{force}$$

$$= 0\cdot75 \times 9100$$

$$= 6825 \; \text{newton metre}$$

Hence, using this result

$$\text{Power transmitted} = \frac{2\pi \times 6825 \times 150}{60}$$

$$= 107\,\mathbf{kW}$$

Some quantities measured with named units in other systems have only the compound units in SI. In SI named units are kept to a minimum and are already internationally acceptable. Examples of such unnamed compound SI units are those of viscosity. In its kinematic aspect viscosity is concerned with the sliding of one layer of a fluid over another. We have, in fact, an area moving over an area and hence the unit of kinematic viscosity is the *square metre per second* (m^2s^{-1}).

For dynamic viscosity, we need to measure the force which slides unit areas over one another in unit time and hence the unit of dynamic viscosity is the *newton second per square metre* (N s m^{-2}).

As an example one may take the expression giving the viscosity of a liquid in terms of its flow under pressure and at constant temperature through a capillary, assuming that the flow is not turbulent.

$$\text{Here viscosity} = \frac{p\pi r^4}{8Vl}$$

If p is in N m, r and l in metre, and V is m^{3}s^{-1} then the viscosity is found in the correct SI form. It must be admitted that here, for a small scale laboratory experiment, the SI units seem at first less convenient than the older cgs measures. One can, however, avoid difficulties by taking linear measures in millimetre, inserting suitable powers of ten or decimal points in the computation. Students will find that the use of a uniform system will avoid arithmetical errors.

Example 14

A light oil of density given as 890 g/litre runs through a horizontal capillary tube of length 820 mm at a rate of 71 ml/min when maintained in a reservoir at a head of 963 mm. What is its viscosity?

Arranging all quantities in the required SI form

$$\text{density} = 890 \, \text{kg} \, \text{m}^{-3}$$

$$\text{head} = 0.963 \, \text{m}$$

$$\text{hence pressure} = 890 \times 0.963 \times 9.81 \, \text{N} \, \text{m}^{-2}$$

$$r = 10^{-3} \, \text{m}$$

$$r^4 = 10^{-12} \, \text{m}$$

$$\text{length} = 0.82 \, \text{m}$$

$$\text{flow} = 1.17 \, \text{ml} \, \text{s}^{-1}$$

$$= 1.17 \times 10^{-6} \, \text{m}^3 \, \text{s}^{-1}$$

$$\text{Then } \frac{p\pi r^4}{8Vl} = \frac{890 \times 0.963 \times 9.81 \times 3.14 \times 10^{-12}}{8 \times 1.17 \times 10^{-6} \times 0.82}$$

$$= \frac{8.9 \times 9.63 \times 9.81 \times 3.14 \times 10^{-11}}{8 \times 1.17 \times 8.2 \times 10^{-7}}$$

$$= 34 \times 10^{-4} \, \text{N} \, \text{s} \, \text{m}^{-2}$$

$$= 3.4 \, \text{mN} \, \text{s} \, \text{m}^{-2}$$

A similar small scale effect is that of surface tension, which in SI would be measured in newton per metre. Calculation, arising say from the height of water in a capillary tube, would be handled as in Example 14.

4

Heat and Thermodynamics

Since heat is a form of energy, an expression of the kinetic energy of the molecules of a substance, it can be measured in joule. Where the heat is produced in the products of combustion in air, it is usual to consider the energy produced from a unit mass of the combustible substance itself, the so called *calorific value*, heat of combustion or specific energy. This is the form of greatest use to the engineer concerned with fuels as a source of heat and power.

Examples of tabulated values are

Pulverized coal	$28 \cdot 3\,\mathrm{MJ\,kg^{-1}}$
Anthracite	$33 \cdot 1\,\mathrm{MJ\,kg^{-1}}$
Coke	$29 \cdot 5\,\mathrm{MJ\,kg^{-1}}$
Kerosene (paraffin)	$44 \cdot 0\,\mathrm{MJ\,kg^{-1}}$

Gaseous fuels are more conveniently measured by volume.

Town gas	$18 \cdot 1\,\mathrm{MJ\,m^{-3}}$
Natural gas	$34 \cdot 3\,\mathrm{MJ\,m^{-3}}$

These are empirical values for the available energy, obtained experimentally. Note that it is convenient to use a decimal point here rather than a smaller multiple.

Example 15

A 15 MW power station using pulverized fuel burns 11·3 Mg/h under full load. What is its overall efficiency?

Using the tabulated value

$$\text{Energy equivalent of fuel used in one hour} = 11{\cdot}3 \times 10^3 . 28{\cdot}3 \times 10^6\,\mathrm{J}$$

$$\text{Hence total power input} = \frac{11{\cdot}3 \times 10^3 \times 28{\cdot}3 \times 10^6}{3600}\,\mathrm{W}$$

$$= 90\,\mathrm{MW}$$

$$\text{Hence efficiency} = \frac{15 \times 100}{90}\,\%$$

$$= 16{\cdot}7\,\%$$

When heat is produced or absorbed by chemical reactions other than large scale combustion, or if the combustion reactions themselves are being studied, the calorific value is no longer an appropriate measure. Since one is then concerned with entities such as molecules or atoms, SI is likely to adopt, as a basic unit of quantity of substance, the *mole* (mol). This reformulates the old 'gramme molecule' already familiar to students of chemistry, but in terms of a particle count. It is not yet (May 1970) formally adopted.

Example 16

Find the energy of the bond C—H in methane, CH_4, given the following experimental data:

(i) The reaction $2H_2 + O_2 = 2H_2O$ releases $288\,kJ\,mol^{-1}$ as heat.

(ii) The reaction $C + O_2 = CO_2$ releases $404\,kJ\,mol^{-1}$ as heat.

(iii) The reaction $CH_4 + 2O_2 = CO_2 + 2H_2O$ releases $885\,kJ\,mol^{-1}$ as heat.

(iv) The bond energy of H_2 is $426\,kJ\,mol^{-1}$.

(v) The mean bond energy of solid carbon is $628\,kJ\,mol^{-1}$.

Assuming that the heat of formation of methane is the difference between that released when it is burnt and that released by its constituent carbon and hydrogen burnt separately, we get

1 mol of carbon C yields	$404\,kJ$
2 mol of hydrogen H yield	$576\,kJ$
	———
	$980\,kJ$
1 mol of CH_4 yields	$885\,kJ$
	———
	$95\,kJ$
	———

To this must be added the bond energy from the molecular carbon and hydrogen from which the CH_4 molecule is produced, $628\,kJ$ for the C and $(2 \times 426)\,kJ$ for the $2H_2$.

Hence, total energy $= (95 + 628 + 852)\,kJ$

$= 1575\,kJ$

Hence, each C—H bond has energy

given by $(1575 \div 4)\,kJ\,mol^{-1}$ $= 394\,kJ\,mol^{-1}$

Examples 15 and 16 have used the joule as a measure of heat energy on the assumption that the measurements can in fact be made. In general, heat energy cannot be measured directly in terms of the accelerated masses which define the derived units of Section 3. One actually measures the *temperature* of a substance, calculating energy changes from temperature changes according to well-known physical laws. SI uses the well established Absolute or Kelvin scale of temperature, taking as zero the (extrapolated) temperature at which a substance has nominally zero heat energy, and as its calibration point the 'triple point' of water, i.e. the temperature (and hence energy level) at which the solid, liquid and vapour phases of water are in equilibrium in a closed evacuated vessel. This temperature interval is not, however, divided into the usual metric subunits, but into 273·16 parts each of which is one *kelvin* (K) or one *degree kelvin*. This makes the kelvin substantially equivalent to the degree Celsius (°C), whose zero is at 273·15 on the Kelvin scale. The name Centigrade is not used since it has a different meaning in many countries (see p. 6).

It follows that one uses the °C for recording and measuring temperatures, but the Kelvin scale for energy computations. For most practical purposes

$$x°C = (x + 273)K$$

The convention followed here is to use K or °C for a measure of temperature, but K for a temperature interval of one degree. Thus the boiling point of water is 100°C or 373K, its specific heat capacity at 15°C is $4190 \, J \, kg^{-1} \, K^{-1}$.

Changes in volume and pressure of gas as the energy content changes are now apparent as changes of temperature. The familiar law for one mole of ideal gas is

$$PV = RT$$

R, the gas constant, has the value

$$8·31 \, J \, mol^{-1} \, K^{-1}$$

Example 17

What mass of CO_2 at 27°C occupies $3 \, m^3$ at a pressure of $10 \, kN \, m^{-2}$?

For $n \, mol$ of gas, assuming that for these values CO_2 behaves as an ideal gas.

$$PV = nRT$$

$$P = 10 \, kN \, m^{-2}$$

$$V = 3 \, m^3$$

$$T = (27 + 273)K$$

$$\text{hence, } n = \frac{10\,000 \times 3}{8·31 \times 300}$$

$$= 12 \, mol$$

$$\text{and mass} = 12 \times 44 \, g$$

$$= 528 \, g$$

For solids and liquids the energy increment corresponding to a given temperature change is usually given in joule per kilogramme degree.

This quantity is the *specific heat capacity.*

Here are typical values, obtained experimentally:

Water	$4190\,\mathrm{J\,kg^{-1}\,K^{-1}}$
Ice	$2100\,\mathrm{J\,kg^{-1}\,K^{-1}}$
Copper	$390\,\mathrm{J\,kg^{-1}\,K^{-1}}$
Steel	$404\,\mathrm{J\,kg^{-1}\,K^{-1}}$

Values are usually measured at 15°C, and may vary considerably with temperature and pressure. Note that using the solidus the unit is written $\mathrm{J/kg\,K}$ and *not* $\mathrm{J/kg/K}$.

Example 18

A litre of boiling water is poured into an unwarmed stainless steel teapot of mass 250 g, initially at a temperature of 15°C. At what temperature does the tea begin to brew?

Taking the tabulated values:

Initial energy level above 0°C $= (4190 \times 100) + (0{\cdot}25 \times 404 \times 15)\,\mathrm{J}$

Final energy level above 0°C $= [4190 + (0{\cdot}25 \times 404)] \times t$ in joule

Where t is the final temperature in degrees.

Hence, assuming no losses have occurred,

$$t = \frac{419\,000 + 1515}{4190 + 101}$$

$$= 98 \text{ to 2 sig. figs.}$$

and final temperature $= 98°C$

Specific latent heats of fusion or vaporization do not involve temperature changes and are given immediately in joule/kilogramme. The actual values, of course, vary considerably with the temperature and pressure at which the change of state occurs.

Example 19

Feed water supply initially at 15°C is used to generate steam at a pressure of $3{\cdot}4\,\mathrm{MN\,m^{-2}}$ absolute and a temperature of 350°C. What is the total energy taken up by the steam?

The total energy is made up of three terms:

(i) Energy to raise water from 15°C to its boiling point at the given pressure.

(ii) Energy to evaporate the water into steam at the same temperature.

(iii) Energy to superheat the steam.

At this range of temperature and pressure changes constant values of specific or latent heats are inappropriate, and one refers to steam tables (which assign zero energy to water at 0°C). These give, for $3.4\,\text{MN}\,\text{m}^{-2}$

Boiling point of water	240°C
Specific heat energy of boiling water	$1050\,\text{kJ}\,\text{kg}^{-1}$
Specific latent heat of steam at 240°C	$1760\,\text{kJ}\,\text{kg}^{-1}$
Specific energy of superheat from 240°C–350°C	$310\,\text{kJ}\,\text{kg}^{-1}$

$$\text{Hence, total energy per kg} = (1050 + 1760 + 310)\,\text{kJ}\,\text{kg}^{-1}$$
$$= 3120\,\text{kJ}\,\text{kg}^{-1}$$
$$\text{Energy per kg of intake water at 15°C} = (15 \times 4190)\,\text{J}\,\text{kg}^{-1}$$
$$= 63\,\text{kJ}\,\text{kg}^{-1}$$
$$\text{Hence, energy per kg taken up} = (3120 - 63)\,\text{kJ}\,\text{kg}^{-1}$$
$$= 3.06\,\textbf{MJ}\,\textbf{kg}^{-1}$$

The conduction of heat is the transfer of energy. Since one is usually concerned with the rate of this transfer the unit chosen will be the watt. For heat insulating laggings the energy is transferred to the atmosphere, and represents a loss. This is proportional to both the internal temperature above the ambient temperature, and the area exposed to cooling.

Hence one measures the *thermal conductivity* of, say, an insulating panel of a given thickness in watt per square metre degree, where $1\,\text{W} \equiv 1\,\text{J}\,\text{s}^{-1}$. For example, the heat loss through window glass $3\,\text{mm}$ thick is approximately

$$6\,\text{W}\,\text{m}^{-2}\,\text{K}^{-1}$$

for a constant temperature gradient.

The temperature here is the difference between the ambient *air* temperatures on each side of the glass, so that the figure is the SI equivalent of the heat engineer's U value. Given the area and tabulated heat loss for structures such as windows, cavity walls and ceiling boards, one can directly obtain the total loss from a room in watt. This is equal to the required output of a heating system to maintain the temperature at a given level above the outside.

The actual thermal conductivity of a substance will be the rate at which energy is transferred through it along a constant temperature gradient, i.e. the number of joule per second for each square metre of area for each metre of thickness for each degree of temperature.

$$\text{Hence, the unit is } 1\,\text{J}\,\text{m}\,\text{s}^{-1}\,\text{m}^{-2}\,\text{K}^{-1} \equiv 1\,\text{W}\,\text{m}^{-1}\,\text{K}^{-1}$$

For copper, the value is about $400\,\text{W}\,\text{m}^{-1}\,\text{K}^{-1}$. These units are not named. The thermal conductivity of a substance is of no great value since in most practical situations, e.g. the transfer of heat from hot flue gases to water through the metal of a boiler tube, the existence of deposits and of stagnant boundary layers of fluid affects the results more than the thickness or nature of the metal. The U value just quoted is an experimental coefficient taking such factors into account.

Finally, one must discuss reversible adiabatic energy changes. If a mass of gas in a heat insulated cylinder is compressed by a piston external work is done but no heat is added. The three parameters pressure, volume and temperature all change: what is considered to remain constant is the *entropy* of the gas. If heat is added or extracted the entropy changes, and by definition this change, for a very small increment of heat dQ, is the ratio

$$\frac{\mathrm{d}Q}{T}$$

where T is the temperature on the absolute scale.

Hence the SI unit for differential entropy changes is the *joule/degree Kelvin*, $\mathrm{J\,K^{-1}}$. Total entropy is obtained by integration in the usual way.

Example 20

A closed vessel contains 1 kg of nitrogen gas at 0°C. The gas is heated to 100°C. What is the change in entropy?

Since only the pressure changes the heat added for each incremental rise in temperature dT is given by

$$\mathrm{d}Q = c_v \cdot \mathrm{d}T$$

where c_v is the specific heat capacity at constant volume. Hence integral of dQ/T between the limits 273K and 373K is given by

$$c_v \ln\frac{373}{273}$$

Over this temperature range c_v is constant to within 5% and has the value $750\,\mathrm{J\,kg^{-1}\,K^{-1}}$ approximately (see tables).

$$\text{Hence change of entropy} = 750 \times \ln(373/273)$$
$$= 750 \times \ln 1\cdot37$$
$$= 750 \times 0\cdot315$$
$$= 236\,\mathrm{J\,K^{-1}}$$

5

Illumination

ALTHOUGH light can also be regarded as a form of energy, the characteristic responses of living organisms to the particular range of electrodynamic waves we call 'light waves' obviously call for a direct method of measuring light intensity. The earliest attempt to lay down a unit of intensity was the definition of a standard candle of a particular size and composition, and SI has merely brought this unit up to date as the *candela* (cd), in terms of the light emitted from platinum at its point of solidification—a very sharply defined temperature.

What we see as the intensity of a light source is only a function of the energy of the source if the wavelength is constant, and indeed all photoactivity varies with wavelength. It follows that the readings of photoelectric lightmeters, the exposure of film emulsions, the signal generated by a television camera and the reaction of the human eye are all difficult to correlate with one another. For the purposes of illumination engineering, however, the candela is a straightforward measure.

Light emitted from a source travels out in all directions, and is said to form a *luminous flux*. The flux passing through a solid angle of one steradian whose vertex is a point source of one candela is said to be one *lumen* (lm). This implies that the emission is uniform in all directions.

Hence $1\,\mathrm{lm} = 1\,\mathrm{cd\,sr}$. It is easily visualized as a cone of light originating in the source, such as would stream from an opaque shell concentric with the source but having a hole in it. Since the complete sphere surrounding the source has a solid angle of $4\pi\,\mathrm{sr}$ (see section 2) the total flux from a source of one candela is

$$4\pi\,\mathrm{lm}$$

This result assumes that no flux is created or lost in the space around the source, and is therefore a *theorem* in physics. It is known as the Gauss theorem and agrees with experimental results.

Example 21

What is the intensity of a $100\,\mathrm{W}$ electric lamp which produces a luminous flux of $13.5\,\mathrm{lm\,W^{-1}}$?

$$\text{Total flux} \quad = 13{\cdot}5 \times 100\,\text{lm}$$

$$= 4\pi \times \text{intensity}$$

$$\text{Hence intensity} = \frac{1350}{4\pi}\,\text{cd}$$

$$= 107\,\textbf{cd}$$

This result gives the mean spherical candle power (MSCP) of the lamp, which is thus regarded as a uniform source.

If a surface emits lights its *luminance* is expressed in candela per square metre. The corresponding expression for the total luminous flux from a flat surface of total intensity I is $\phi = \pi I$ (see Lambert's theorem and its consequences). The unit of *illumination* is the *lux* (lx) which is that produced by a flux of one lumen falling on one square metre.

Typical values are:

$$\text{Daylight, bright sunshine} \qquad \text{about } 50\,\text{klx}$$

$$\text{Vertical tropical sunshine} \qquad \text{about } 100\,\text{klx}$$

Example 22

A uniform source of intensity 250 cd is mounted 10 m normally from a white panel 500 mm $\times$ 700 mm which reflects 80% of the incident light. What is the total intensity of the light from the panel?

For the dimensions given all parts of the panel can be taken as equidistant from the light.

1 sr intercepts an area of 100 m^2 on a sphere of radius 10 m surrounding the source. Hence illumination in region of panel is $\frac{250}{100}\,\text{lm}\,\text{m}^{-2}$. (Note that this conclusion from the Gauss theorem corresponds exactly to the inverse square law: the two are equivalent.)

$$\text{Hence flux reflected} = \frac{250 \times 0{\cdot}5 \times 0{\cdot}7 \times 80}{100 \times 100}\,\text{lm}$$

$$= 0{\cdot}7\,\text{lm}$$

and, from Lambert's theorem

$$\text{Intensity} = \frac{0{\cdot}7}{\pi}\,\text{cd}$$

$$= 0{\cdot}22\,\textbf{cd}$$

6

Electrical Phenomena

THE system of electrical measurements incorporated in SI was proposed at the turn of the century, internationally agreed some thirty years later, and has been used almost exclusively in technical education since the end of the Second World War. This section, then, is addressed mainly to the science student studying physics at school. The student of electrotechnology will already be familiar with the units discussed.

Physics teaching, over and above any complexities of the subject matter, has had to cope with alternative and coexistent systems of units: often a whole chapter in a textbook of electricity would be devoted to these. SI enables each unit to be defined as the need for it arises in developing the theory, and calls for no cross-conversions or alternative measures.

Electricity, unlike all the other quantities proposed as fundamental by SI, does not correspond to direct intuition from a sensory response. It is known only by the phenomena it produces. To be able to measure these implies some knowledge of them, so that this section can scarcely avoid the appearance of a potted textbook. I must, however, repeat that the aim of this work is to introduce the *units*: it is assumed that the concepts are known and are being studied in forms quantified by systems other than SI.

The SI electrical units are also known as the rationalized MKS or Giorgi system, although originally these latter used a slightly different starting point.

Whatever may have been the history of the subject, it is clear that there is one electrical phenomenon of overriding technical importance: the electric current. It is this that travels through the complex network of wires and cables that enmeshes our homes, our factories, our streets and our countryside; although, of course, what we actually pay for is the *energy* it transmits.

SI, then, begins with a definition which will be given in full (see bibliography).

The *ampere* (A) is

> the constant current which, if maintained in two straight parallel conductors of infinite length, of negligible circular cross section and placed at a distance of one metre apart in a vacuum, will produce between them a force equal to 2×10^{-7} newton per metre of their length.

The value 2×10^{-7} makes the SI ampere correspond exactly in magnitude with that of an earlier definition. To choose 1 N for the force would have produced a new fundamental unit, counter to the principles of SI (see p. 1).

Example 23

Two long straight wires 2 mm apart carry currents of 3 A and 2 A. What is the force between them, given that it is proportional to the currents and inversely proportional to the distance between the wires?

$$\text{From the data } F = K \cdot \frac{I_1 I_2}{r}.$$

K is a constant, and, by definition, $K = 2 \times 10^{-7}$ if the other quantities are unity, hence K takes this value

$$\text{and } F = \frac{2 \times 10^{-7} \times 3 \times 2}{0 \cdot 002}$$

$$= 6 \times 10 \, \text{Nm}^{-1}$$

The constant K above will need further discussion later.

It will be convenient to consider electrical phenomena under three sub-headings:

1 Current phenomena

SI itself makes no assumptions about the nature of a current, it merely measures its behaviour. If a steady current of 1 A is allowed to flow for one second we have, presumably, a *quantity* or, as it is usually called, a *charge* of electricity. The unit of charge is thus the *ampere second* or *coulomb* (C). The charges of accumulators are usually given in ampere hours (A h).

When a current is driven around a circuit by a battery or a generator work is done, often merely in raising the temperature of the conductor, which afterwards dissipates its energy by convection or radiation. The rate at which work is done in watts measures the *electromotive force* (e.m.f.) of the source of electrical energy.

SI defines the *volt* (V) as that constant e.m.f. which applied to the ends of a conductor produces in it a steady current of one ampere when the power dissipated is one watt.

$$1 \, \text{V} \equiv 1 \, \text{W A}^{-1}$$

Notice that the definition does not imply that the power *is* dissipated. In a transformer, for example, the input energy is transferred to the field energy of the core. Hence one can speak of an electric fire using 8 A at 250 V as a 2 kW fire, but a transformer with this input would be rated as 2 kVA.

The current driven by a constant e.m.f. does not increase indefinitely (compare the acceleration of a mass under a constant force) since it encounters resisting forces explicable in terms of electron scatter at crystal lattice irregularities. The work which eventually appears as heat is done in overcoming these. SI defines unit *resistance* as that of a conductor in which an applied e.m.f. of one volt produces a current of one ampere. The unit is the *ohm* (Ω).

Both definitions assume that the only source of e.m.f. is external to the conductor. These units will be familiar to the student already, but one example may prove helpful, and is given to make the account complete.

Example 24

A secondary cell produces an e.m.f. of 1·95 V when connected to the ends of a wire 10 m long of resistance $2\,\Omega\,\mathrm{m}^{-1}$. Calculate the current produced, the power dissipated and the quantity of electricity discharged from the cell after 10 hours.

$$\text{Resistance} \quad = 20\,\Omega$$

$$\text{Hence current} = \frac{1\cdot95}{20}\,\mathrm{A}$$

$$= 98\,\mathrm{mA}$$

$$\text{Power} \quad = 0\cdot098 \times 1\cdot95\,\mathrm{W}$$

$$= 191\,\mathrm{mW}$$

$$\text{Discharge} \quad = 0\cdot98\,\mathrm{Ah}$$

$$= 0\cdot98 \times 3600\,\mathrm{C}$$

$$= 3530\,\mathrm{C}$$

In SI units the resistivity of a given substance is, to be consistent, the resistance between opposite faces of a sample 1 m long and of cross section 1 m², and hence is expressed in ohm metre squared per metre, i.e. in ohm metre. In practice, however, tables are available giving the resistance in ohm per metre of standard conductors in all available cross sections and metals.

A further important property of a current is that of electrolytic action. Students will realize that a current through an electrolyte differs in kind from that through a wire, since actual transfer of matter from one electrode to the other takes place, producing products of electrolysis. SI expresses the standard result that, during electrolysis

One gramme-equivalent of an element is liberated by 96 500 coulomb without attempting to name this quantity.

Example 25

Using platinum electrodes $0.5\,A$ is passed for $20\,min$ through very dilute sulphuric acid. What volume of oxygen is evolved?

$$\text{Quantity of electricity} = 0.5 \times 20 \times 60$$
$$= 600\,C$$

$$\text{Hence mass of oxygen liberated} = \frac{600}{96\,500}\,\text{gramme-equiv.}$$
$$= \frac{8 \times 6}{965}\,g$$
$$= \frac{8 \times 6}{965 \times 32}\,\text{mol of}\,O_2$$

and, since $1\,mol$ occupies $22.4\,l$ at S.T.P.

$$\text{Volume of oxygen} = \frac{6 \times 22.4}{965 \times 4}\,l$$
$$= 34.8\,ml\,\text{at S.T.P.}$$

II Electrostatic Field Phenomena

It is at this point that some of the SI units differ sharply from those which may already be familiar to science students. The static unit of charge in SI is the coulomb, already defined as that equivalent to a current flow of one ampere second. Thus the charge on an electron is given in SI units as $1.6 \times 10^{-19}\,C$ and is of course negative. The proton has the corresponding positive charge. Since electrostatic forces between charges obey the inverse square law the field of force surrounding a point charge can be expressed using the Gauss theorem. The term 'electric flux' is used by analogy with luminous flux, but is rather misleading. There is no 'cone of influence' produced if the source is surrounded by a hollow shell with a hole cut in it (see page 20). In these circumstances the charge induces an equal charge distributed over the surface of the shell. The total flux from a point source, then, is measured in *coulomb* and *not* coulomb steradian: the flux from a source, in SI, is taken as equal to the charge.

When flux passes through a surface, imagined as drawn or projected on a sphere surrounding the charge, we get the measure of *electric flux density*, whose unit is the coulomb per square metre. This quantity is usually symbolized by D, or by the vector $\mathbf{D}$ if its direction from the source as origin is to be considered.

Example 26

What is the value of D at a distance of 10 mm from a very small body having a charge of 3 pC?

$$\text{Total flux} = 3\,\text{pC}$$

$$\text{Area of sphere} = 4\pi \times 100\,\text{mm}^2$$

$$\text{Hence electric flux density} = \frac{3}{4\pi \times 100}\,\text{pC}\,\text{mm}^{-2}$$

$$= \frac{3 \times 10^{-12}}{4\pi \times 10^{-4}}\,\text{C}\,\text{m}^{-2}$$

$$= 2{\cdot}38 \times 10^{-7}\,\text{C}\,\text{m}^{-2}$$

Note that if the charge were actually distributed over a sphere of radius 10 mm the value of D very close to its surface would also be $2{\cdot}38 \times 10^{-7}$ $\text{C}\,\text{m}^{-2}$ and normal to it. Note also that D is independent of the medium surrounding the charge.

The *potential difference*, the work done in moving a charge between two points of an electrostatic field that exerts a force on it, is measured in SI by joule per coulomb. That is, unit potential difference between two points is defined when one joule of work is done in moving a coulomb of charge from one point to the other. The student will know that this work is independent of the path taken.

One sees easily that if

$$\text{unit of p.d.} \equiv \text{joule/coulomb}$$

this can be written as

$$J/C = \frac{\text{kg}\,\text{m}^2}{\text{A}\,\text{s}^3}$$

$$= \text{W}\,\text{A}^{-1}$$

$$= \text{V}$$

Hence in SI as in other systems the unit of potential difference is the same as that of electromotive force. An e.m.f. can be thought of as maintaining a potential difference between the ends of a conducting wire, so that the charges constituting a current move along it. A difference in potential between two points in an electrostatic field not equidistant from the source implies a *potential gradient* between them. Clearly the SI measure for this is in *volt per metre*.

It can be shown that the force on a unit charge at any point in a field is equal to the instantaneous value of the potential gradient at the point, and is in the same direction as the vector $\boldsymbol{D}$. This gives the vector $\boldsymbol{E}$, the *electric field strength*, whose magnitude is measured in newton per coulomb. It is interesting to note that SI shows this equivalence at once.

For potential gradient

$$V/m \equiv W A^{-1} m^{-1}$$

$$\equiv \frac{kg\,m}{A\,s^3}$$

For E, the electric field strength

$$N/C \equiv N A^{-1} s^{-1}$$

$$\equiv \frac{kg\,m}{A\,s^3}$$

Students familiar with other systems may find it interesting to consider the ratio D/E

The units of D are	$C m^{-2} \equiv A s\,m^{-2}$
The units of E are	$N C^{-1} \equiv kg\,m A^{-1} s^{-3}$
Hence the units of D/E are	$A^2 s^4 kg^{-1} m^{-3}$

This rather portentous expression, which will be simplified later, gives us the measure of the physical quantity known as *permittivity*. In SI its numerical value for free space is

$$\epsilon_0 = 8 \cdot 85 \times 10^{-12} \text{ units}$$

There is further discussion in Example 32.

Students are sometimes puzzled to find that some text books using systems of units other than SI state that *in vacuo* $\boldsymbol{D}$ and $\boldsymbol{E}$ are identical, which appears to be a contradiction if the ratio of their magnitudes has a measure. The answer is quite simply that $\boldsymbol{D}$ and $\boldsymbol{E}$ do not have a tangible physical existence: they are merely vectors used to describe mathematically the observed properties of the field. The different systems describe the field in different ways; but *what* they describe is unknown and possibly unknowable.

The *relative permittivity* is then the ratio between the permittivity of a medium and that of free space, and this, of course, is independent of the units used. It is nearly unity for air.

Example 27

What is the field strength at a point 100 m from the charge in Example 26?

$$\text{Since } D = \frac{Q}{4\pi r^2}$$

$$\text{and } E = D/\epsilon_0$$

$$\text{we get } E = \frac{3 \times 10^{-12}}{4\pi \times 8 \cdot 85 \times 10^{-12} \times 10^4}$$

$$= 2 \cdot 69 \times 10^{-6} N C^{-1}$$

Since Newton's third law applies this is also the force on the $3\,\mathrm{pC}$ charge: the expression:

$$\frac{Q_1 Q_2}{4\pi\epsilon_0\, r^2}$$

gives the force between charges directly in newton if Q is in coulomb and r in metre.

Finally, the *capacitance* of a capacitor or a body capable of taking a charge is measured in *coulomb per volt*. For a capacitor this voltage is the potential difference between its plates, and its capacitance is the charge required to establish this difference. This, in SI, becomes the named unit *farad* (F).

$$1 \text{ farad} = 1 \text{ ampere second per volt}$$

This unit will already be familiar. Its common submultiples are the μF and pF: the farad itself is very large.

$$\text{Since } \mathrm{C V^{-1}} \equiv \mathrm{A s V^{-1}}$$

$$\equiv \frac{\mathrm{A^2 s}}{\mathrm{W}}$$

$$\equiv \frac{\mathrm{A^2 s^4}}{\mathrm{kg\, m^2}}$$

and since the ratio $\dfrac{D}{E} \equiv \dfrac{\mathrm{A^2 s^4}}{\mathrm{kg\, m^3}}$ (see page 27)

it follows that the unit of permittivity already mentioned is in SI equivalent to *farad per metre*. Since D is given in units per square metre and E in units per metre, a more satisfactory measure of ϵ_0 would be in *farad metre per square metre* which expresses more clearly the field effects between capacitor plates of given area and separation. The simpler name is, however, always used.

Example 28

A capacitor has two sets of plates each of effective size 200 mm $\times$ 150 mm separated by mica $0{\cdot}25$ mm thick and of relative permittivity 4. What is its capacitance if edge effects are neglected?

The capacitance is given by $\dfrac{\epsilon A}{d}$

$$\text{here } \epsilon = 4 \times \epsilon_0$$

$$= 4 \times 8{\cdot}85 \times 10^{-12}$$

$$A = 0{\cdot}2 \times 0{\cdot}15\,\mathrm{m^2}$$

$$d = 0{\cdot}00025\,\mathrm{m}$$

$$\text{and capacitance} = \frac{4 \times 8 \cdot 85 \times 10^{-12} \times 0 \cdot 03}{2 \cdot 5 \times 10^{-4}} \text{ farad}$$

$$= 4 \cdot 2 \times 10^{-10} \, \text{F}$$

$$= 420 \, \text{pF}$$

Finally the student must recognize that although we now regard a current as a drift (or perhaps a sequential exchange) of negatively charged particles through a conductor, for purposes of measurement a current rather than a charge is chosen as unit. Physically, the charge of the particle is fundamental: for the practical conveniences of measurement it is treated as derivative.

III *Electrodynamic field phenomena*

Although a wire carrying a current is electrostatically neutral, when *two* parallel wires carry currents in the same direction the negative charges in one are moving relative to the positively charged atoms at rest in the second wire, and, symmetrically, the positive charges are moving relative to the drifting electrons. Relativity theory shows that in these circumstances the wires acquire equal and opposite charges *relative to one another*, although they remain neutral to external measurement. Hence the wires attract. Similarly, the wires repel if the currents are in opposite senses.

This is, incidentally, an exceptional example of a relativistic effect produced by velocities small compared with that of light. The force appears because the attraction or repulsion between the charges is extremely large. Normally, in electrostatics, we only handle extremely small charges.

The force between the wires is proportional to the relative drift velocities, and hence to the current. The effects of this interaction, considered as a field of force, gives us a new set of phenomena, those known as electrodynamic or more usually as electromagnetic. SI electrodynamic units derive from the proportionality of the effects to the current: the ampere remains the fundamental unit.

The first electromagnetic unit, then, is of *magnetomotive force*, designated by F and measured in *ampere*. We may think of the m.m.f. being responsible for the electromagnetic field effects as the e.m.f., in volt, is responsible for current effects. The textbooks usually refer to a circuit carrying 1 A as an *ampere-turn* and state the m.m.f. in this form, i.e. as 1 AT. Similarly the m.m.f. of a coil of N turns carrying a current I is given as NI ampere-turns. As, however, this is equivalent to N separate circuits each carrying I ampere, or, alternatively, as one circuit of NI ampere, the word 'turn' is not logically required and is omitted in SI. Students are advised to retain it in thinking of such a situation, but to omit it in recording results.

Corresponding to the vector E in the static field, measured in volt per metre, we have *magnetic field strength*, the vector H, in *ampere per metre* in a direction given by the familiar Left Hand Rule. This is easily visualized by considering the following:

Example 29

2000 turns of fine wire are wound evenly on a core of very small radius over a length of 150 mm. The wire carries a current of 2 mA. Find the m.m.f. applied to the core, and the value of the magnetic field strength within the winding. Neglect end effects.

$$\text{Total m.m.f.} = 2000 \times 2\,\text{mA}$$

$$= 4\,\text{A}$$

$$\text{Magnetic field strength} = \frac{4}{0\cdot15}\,\text{A m}^{-1}$$

$$= 26\cdot7\ \textbf{A m}^{-1}$$

In practice, of course, the end effects are important. The next unit needs care in formulation. The electric field strength E is also measured (see page 26) as the force in newton acting on a charge of one coulomb. There is nothing corresponding to the coulomb in the electrodynamic field, so one measures instead the force acting per metre length on a wire carrying a current of one ampere. Since we are considering the force per unit length, we are, however, measuring the *density* of a field effect. By analogy, this is called the *magnetic flux density B*, or, if direction is implied the magnitude of the vector B, in newton per ampere metre. This unit is named the *tesla* (T), although the name has not yet reached general use in technical literature. Note that, if the tesla is simplified—

$$T \equiv N\,A^{-1}\,m^{-1}$$

$$\equiv \frac{\text{kg m}}{\text{s}^2\,\text{A m}}$$

If it is then multiplied both top and bottom by the unit m s

$$T \equiv \frac{\text{kg m}^2\,\text{s}}{\text{s}^3\,\text{A m}^2}$$

$$\equiv \frac{\text{W s}}{\text{A m}^2}$$

$$\equiv \text{V s m}^{-2}$$

It is now in a form corresponding to the electrostatic vector D. The magnitude of D was measured in ampere second per square metre; that of B is in volt second per square metre. The ampere second or coulomb was the unit of electric flux, so called, and the volt second is taken as the unit of magnetic flux. It is named the *weber* (Wb), so that an alternative expression for B is in weber per square metre.

Another definition of the weber is possible, by considering not the absolute value of the flux, but its rate of change. If it is reduced at a uniform rate from a given value to zero in one second, and in so doing induces an e.m.f. of one volt in a single loop linked with the changing flux, then this initial value defines the weber. The student will be aware that for the magnetic field, the total flux around a point in free space is zero.

Example 30

A measuring instrument consists of a cylindrical soft iron core of height and diameter both 40 mm which can rotate between the pole pieces of a permanent magnet so that the flux density is everywhere radial and of value $2\,\mathrm{mW\,m^{-2}}$. A square coil of 50 turns is wound on the core and carries the input current. What is the torque when the current is 2 mA?

$$\text{Torque} = 2\,N\,Brl\,I$$
$$\text{and here } N = 50$$
$$r = 20\,\mathrm{mm}$$
$$l = 40\,\mathrm{mm}$$
$$B = 2\,\mathrm{mWb\,m^{-2}}$$
$$I = 2\,\mathrm{mA}$$
$$\text{Hence, torque} = 2 \times 50 \times 2 \times 10^{-3} \times 2 \times 10^{-2} \times 4 \times 10^{-2} \times 2 \times 10^{-3}$$
$$= 3{\cdot}2 \times 10^{-7} \text{ newton metre}$$

The last named unit to be considered, that of inductance, is the henry, and is already in general use: in fact students are often unaware of the two c.g.s. equivalents. In SI it is seen to correspond exactly to the farad. For a capacitor, the build up of charge as current flows in produces an energy distribution whose total value increases until the p.d. between the plates has reached its maximum. The charge energy is proportional to the voltage and hence capacity is measured in ampere second per volt. For an inductance, the build up of current in the coil produces an energy distribution whose total increases until the current has reached its maximum. The magnetic field energy is proportional to the current and is measured in weber (or volt second) per ampere. This unit is the *henry* (H).

Finally, the ratio B/H in electrodynamics corresponds to D/E in electrostatics, and has a numerical value in SI different from that in other systems. Note that D depends on the charge which is the source of the field while E depends on the medium; in the electrodynamic field it is H that depends solely on the current maintaining it, while B depends on the medium.

B is measured in volt second per square metre.

H is measured in ampere per metre.

Hence the unit of the ratio is

$$\frac{B}{H} \equiv \frac{\mathrm{V\,s}}{\mathrm{m}^2} \times \frac{\mathrm{m}}{\mathrm{A}}$$

$$\equiv \mathrm{V\,s\,A^{-1}\,m^{-1}}$$

$$\equiv \mathrm{H\,m^{-1}}$$

that is, in henry per metre, corresponding to the farad per metre for permittivity. The ratio is the quantity *permeability* symbolized as μ.

In SI, the permeability of free space is

$$\mu_0 = 4\pi \times 10^{-7}\,\mathrm{H\,m^{-1}}$$

The permeability of most materials is approximately this value, hence their *relative permeability* is about unity. For the very exceptional but important group of ferromagnetic materials, the relative permeability may be as high as 10^5.

It is now worth while reconsidering Example 23, a clear understanding of which is the key to appreciating the basis of SI electrical measures.

The force between unit length of two parallel long wires was given by

$$K \cdot \frac{I_1 I_2}{r}$$

A better expression is

$$F = \frac{\mu_0 I_1 I_2}{2\pi r}$$

here 2π is a numerical constant arising from the geometry of the field, F, I_1, I_2, and r are in any suitable units, and μ_0 is a quantity whose value and dimensions depends on the units chosen. Since we *define* unit current $I_1 = I_2$ as one ampere if r is one metre *in vacuo* and if F is 2×10^{-7} newton it follows that μ_0 *must be* $4\pi \times 10^{-7}\,\mathrm{H\,m^{-1}}$ for free space in this system. Students who have met them may remember that the Heaviside–Lorentz units begin by defining the value of μ_0 and hence deriving the unit of current; but SI does it the other way round.

Since Maxwell's theory shows that for all units (except the Gaussian which some students may also have met)

$$\frac{1}{\epsilon_0\,\mu_0} = c^2$$

where c is the velocity of light known to be

$$3 \times 10^8 \,\mathrm{m\,s^{-1}}$$

it follows that the numerical value of ϵ_0 is

$$8.85 \times 10^{-12}$$

as given on page 27.

The final example is a calculation illustrating several SI units, with the value of μ for air taken as equal to μ_0.

Example 31

What is the inductance of 1000 turns of wire wound on a cardboard former 400 mm long and of radius 5 mm?

The energy stored in the field when the current is I is given by $\frac{1}{2}LI^2$ joule, where L is the inductance.

If we assume that the energy is uniformly distributed and entirely within the coil its uniform density is given by

$$\tfrac{1}{2}\mu_0 H^2$$

The units in this expression will be $\mathrm{J\,m^{-3}}$.

For this coil $H = \dfrac{1000\,I}{0.4}$ in the units $\mathrm{A\,m^{-1}}$

and volume $v = 3.14 \times 10^{-5} \,\mathrm{m^3}$

Hence total energy is

$$\frac{\mu_0 \times 10 \times 3.14\,I^2}{2 \times 0.16} \text{ in joule}$$

and if this is equated to $\frac{1}{2}LI^2$

$$L = \frac{4\pi \times 10^{-7} \times 10 \times 3.14}{0.16} \text{ henry}$$

$$= 250\mu\mathbf{H}$$

The student has now met all the SI electrical units except those defined as reciprocals, and he must learn to think in terms of these rather than in any others he may have met in the past. The symmetry of the electrostatic and electrodynamic units is a great aid both to memory and understanding. The following table shows that interchange of ampere with volt in the derived unit moves us from the static to the dynamic field.

Static		Dynamic	
Quantity	*Unit*	*Quantity*	*Unit*
flux	$\mathrm{A\,s}$	flux	$\mathrm{V\,s}$
D	$\mathrm{A\,s\,m^{-2}}$	B	$\mathrm{V\,s\,m^{-2}}$
e.m.f. (or p.d.)	V	m.m.f.	A
E (as gradient)	$\mathrm{V\,m^{-1}}$	H	$\mathrm{A\,m^{-1}}$
capacitance	$\mathrm{A\,s\,V^{-1}}$	inductance	$\mathrm{V\,s\,A^{-1}}$
permittivity	$\mathrm{A\,s\,V^{-1}\,m^{-1}}$	permeability	$\mathrm{V\,s\,A^{-1}\,m^{-1}}$

Appendix A—A note on dimensions

MANY textbooks of physics and applied mathematics contain chapters on Dimensions. In these chapters units are analysed in terms of mass, length, time, temperature and several possible fundamental electrical quantities. Useful information about the relationships between physical quantities can often be seen in this way, and mistakes in formulation can be avoided.

A side effect of SI so far little publicized is that much of this analysis is contained entirely within the system and does not call for separate treatment. The work on dimensions usually proceeds by ignoring the actual units of measurement and stipulating $\mathbf{M}$, $\mathbf{L}$, $\mathbf{T}$ as notional units. For example, horsepower is analysed as

$$\mathbf{ML^2T^{-3}}$$

ignoring the numerical coefficients that may appear in an actual determination. It should be quite clear that as SI has one and one unit only for each of these fundamental quantities, and builds up its derived units without numerical coefficients (such as the 550 or 33 000 that appears in horsepower calculations) any derived unit is analysed by writing it down without using the shorthand of named derivatives.

Thus, in SI, power is measured by

$$\mathrm{kg\,m^2 s^{-3}}$$

and we call this unit a watt or express it as $\mathrm{J\,s^{-1}}$ merely for convenience. As written, it needs no further analysis.

The last section has used this device freely, to show, for example, the equivalence of potential gradient and electric field strength or the nature of the quantity D/E.

The student will profit by checking quantities measured in SI against the analyses of the chapters on dimensions referred to. One further example only will be given.

Example 32

Establish the dimensional relationship of the quantities permittivity and permeability.

The unit of ϵ_0 is $A\,s\,V^{-1}\,m^{-1}$

The unit of μ_0 is $V\,s\,A^{-1}\,m^{-1}$

Hence the unit of $\epsilon_0\,\mu_0$ is $\dfrac{A\,s\,.\,V\,s}{V\,m\,.\,A\,m} \equiv \dfrac{s^2}{m^2}$

Hence unit of $\dfrac{1}{\epsilon_0\,\mu_0} \equiv \dfrac{m^2}{s^2}$

which is the square of a velocity. Dimensional analysis, of course, cannot determine a numerical value for this result. The value of ϵ_0 can be experimentally determined by considering the force between the plates of a condenser, however, and μ_0 is known (see page 32). The velocity thus found is $3 \times 10^8\,m\,s^{-1}$, which is that of light. If this velocity is determined independently, ϵ_0 can be calculated as on p. 33.

The following two tables analyse, using the fundamental units, all quantities allotted named units in SI, and a few typical quantities measured in derived units. The analyses are transformed into traditional analyses by substituting **M, L, T, I, θ** for kg, m, s, A, K.

Dimensional analysis of typical quantities in terms of
fundamental units

Quantity	*Unit*	*Analysis*
velocity	$m\,s^{-1}$	$m\,s^{-1}$
acceleration	$m\,s^{-2}$	$m\,s^{-2}$
angular velocity	$rad\,s^{-1}$	s^{-1}
volume	m^3	m^3
density	$kg\,m^{-3}$	$kg\,m^{-3}$
viscosity (dynamic)	$N\,s\,m^{-2}$	$kg\,m^{-1}\,s^{-1}$
surface tension	$N\,m^{-1}$	$kg\,s^{-2}$
thermal capacity	$J\,kg^{-1}\,K^{-1}$	$m^2\,s^{-2}\,K^{-1}$
entropy	$J\,K^{-1}$	$kg\,m^2\,s^{-2}\,K^{-1}$
permittivity	$F\,m^{-1}$	$kg^{-1}\,m^{-3}\,s^4\,A^2$
permeability	$H\,m^{-1}$	$kg\,m\,s^{-2}\,A^{-2}$
pressure	$N\,m^{-2}$	$kg\,m^{-1}\,s^{-2}$

Dimensional analysis of quantities having named units in terms of fundamental units

Quantity	Symbol	Unit	Symbol	Definition	Equivalents	Analysis
force	F	newton	N	$kg\,m\,s^{-2}$		$kg\,m\,s^{-2}$
work, energy, heat	W	joule	J	Nm	Ws	$kg\,m^2\,s^{-2}$
power	P	watt	W	$J\,s^{-1}$	$N\,m\,s^{-1}$	$kg\,m^2\,s^{-3}$
luminous flux	Q	lumen	lm	$cd\,sr$		cd
illumination	E	lux	lx	$lm\,m^{-2}$	$cd\,sr\,m^{-2}$	$m^{-2}cd$
electromotive force, potential difference	V	volt	V	$W\,A^{-1}$	$J\,s^{-1}A^{-1},\ N\,m\,s^{-1}A^{-1}$	$kg\,m^2\,s^{-3}\,A^{-1}$
resistance	R	ohm	Ω	$V\,A^{-1}$	$W\,A^{-2}$	$kg\,m^2\,s^{-3}\,A^{-2}$
charge	Q	coulomb	C	As		$s\,A$
capacitance	C	farad	F	$C\,V^{-1}$	$As\,V^{-1}$	$kg^{-1}\,m^{-2}\,s^4\,A^2$
magnetic flux	ϕ	weber	Wb	Vs		$kg\,m^2\,s^{-2}\,A^{-1}$
magnetic flux density	B	tesla	T	$N\,A^{-1}\,m^{-1}$	$Wb\,m^{-2},\ V\,s\,m^{-2}$	$kg\,s^{-2}\,A^{-1}$
inductance	L	henry	H	$Wb\,A^{-1}$	$V\,s\,A^{-1}$	$kg\,m^2\,s^{-2}\,A^{-2}$

Appendix B—Measurement in Atomic Physics

ONE reason for basing SI units on the metre and kilogramme rather than centimetre and gramme is that the units thus defined and derived are of the order of practical magnitudes. The cgs dyne is below the threshold of human discrimination but the newton, as Lighthill* vividly exemplifies it, is the force with which the earth attracts a medium sized apple. It is only to be expected that a few measures—that of surface tension for example—should appear more amenable in cgs units, but on balance the advantages of a uniform system are considerable. This is particularly true at student level where three figure calculations are adequate and indeed less misleading for most purposes. Any adjustment can always be made by inserting suitable powers of 10.

At the level of atomic physics, cgs units are as unwieldy as MKS, and the very structure of atomic physics suggests an internal system of units. Since all charges are multiples of the charge of an electron it is obviously convenient to express an atomic charge in this unit rather than in coulomb. The following list of special units already in international use gives the SI equivalent, and is followed by a few comments. Clearly, a student competent to study atomic physics is not likely to need introduction to its units, but this section is added to complete the

Quantity	Unit	SI equivalent
charge	charge on electron (e^-)	$1{\cdot}6 \times 10^{-19}\,\text{C}$
energy or work	electron volt (eV)	$1{\cdot}6 \times 10^{-19}\,\text{J}$
particle mass	energy equivalent (MeV)	$1{\cdot}8 \times 10^{-30}\,\text{kg}$
velocity	light *in vacuo* (c)	$3 \times 10^{8}\,\text{m s}^{-1}$
action angular momentum	Planck's constant $\left(\dfrac{h}{2\pi}\right)$	$1 \times 10^{-34}\,\text{kg m}^2\text{s}^{-1}$

* Lighthill, M. J. Metrication in Science and Technology (essay in Kellaway, F. W. (ed.), *Metrication*. London, 1968)

list and show how SI can be extended to meet the very special needs of a science where ordinary measures are inappropriate.

The electron volt is the work done, or energy acquired, when a charge equal to that of an electron moves through a p.d. of one volt.

It is now generally known that energy and mass are interchangeable in direct proportion in certain reactions such as electron–positron annihilation. It is convenient then to use one unit rather than two, and since the physicist can measure particle energy more easily than mass the megaelectron volt MeV is taken as a unit. The 'rest' mass of an electron is about $0.5\,\text{MeV}$, obviously more convenient than 9×10^{-31} kg.

The last unit arises from elementary quantum theory. Students who have met this will know that energy transferred when light of frequency n interacts with a particle is always given by

$$W = Nhn$$

where N is an integer and h is a constant (Planck's constant). The quantum of energy corresponding to a given value of n if $N = 1$ is thus

$$W = hn$$

$$= \frac{h\omega}{2\pi}$$

if ω is the angular velocity of the generating circular motion, taking n as frequency of a simple harmonic propagation. Since 2π is merely a constant, the units in which $h/2\pi$ is expressed is seen by dimensional analysis (see Appendix A).

$$\text{joules} \equiv \text{unit} \times \omega$$

$$\equiv \frac{\text{unit}.\text{radians}}{\text{second}}$$

since the radian is dimensionless (see page 6) it follows that the unit of $h/2\pi$ is the joule second or kilogramme metre squared per second. This quantity has the dimensions of angular momentum or 'spin', although the quantity (work $\times$ time) is often called 'action'. In physics sub atomic particles are taken to have integral or half integral spins in terms of this unit.

Appendix C—Measurement in Acoustics

SI does not introduce new named units into acoustics, although it alters a term in a commonly used formula.

A sound source produces energy, at a rate which can be measured in watts. Most people are now familiar with the outputs of amplifiers and other sound equipment so expressed. The *intensity* of a sound is measured as the energy transmitted by the sound waves through unit area in unit time, taking the area as normal to the direction of propagation. Hence, in SI, intensity is measured in watt per square metre, Wm^{-2}. In these units threshold audibility has an intensity of $1\,pWm^{-2}$, while above about $100\,Wm^{-2}$ the ear suffers physical damage.

When sound waves strike a barrier their energy may be

(*a*) reflected as sound

(*b*) absorbed as heat

(*c*) transmitted as sound.

The proportion of energy not reflected, which is lost either by absorption or transmission, gives the constant known as the *sound absorption coefficient*. It is specific to a given material only at a given frequency. Thus, for window glass, it is 0·2 at 500 Hz but drops to 0·06 at 2 kHz.

The reflected sound in a closed hall is reflected repeatedly to produce a reverberation. The *reverberation time* is taken, empirically, as that in which a sound initially of intensity $1\,Wm^{-2}$ decays to threshold audibility at $1\,pWm^{-2}$. The original empirical formula, due to Sabine, was

$$T = \frac{V}{20\,A}$$

where V is the volume of the hall and A is the 'total absorption', i.e. the sum of all surface areas in the room multiplied by their absorption coefficients at the test frequency. Sabine took all linear measurements in feet, choosing 20 as a round figure to give a result agreeing with trials. Clearly, the 20 is a physical quantity, having the dimensions of a velocity, feet/sec. Hence in SI it must be replaced by $\mathrm{m\,s^{-1}}$, and the formula becomes

$$T = \frac{V}{6 \cdot 10\, A}$$

or, in practice

$$T = \frac{V}{6\, A}$$

where all linear measurements are in metre.

The unit of loudness, the phon, is a subjective measure obtained by matching the loudness of a given sound with that of a standard pure tone. It is the logarithm of the ratio of actual intensities and hence is independent of the units in which they are measured. These will now be in watt per square metre.

Similarly, changes in intensity levels, such as arise in considering amplification or sound insulation, are measured, as students will be aware, in *decibel*, again independent of units. By definition, the gain or loss is given by

$$10 \lg \frac{I}{I_0}$$

where I_0 and I are the initial and final levels. In SI the arbitrary standard level against which sound intensities are measured is $1\,\mathrm{pW\,m^{-2}}$, the subjective threshold intensity. The root-mean pressure corresponding to this is $204 \times 10^{-7}\,\mathrm{N\,m^{-2}}$.

Example 33

An empty rectangular hall with a wooden floor and plastered walls and ceiling is $30\,\mathrm{m} \times 10\,\mathrm{m} \times 5\,\mathrm{m}$, and has glazed windows and doors totalling $70\,\mathrm{m^2}$. Find the reverberation time for middle C (256 Hz).

From tables, the absorption coefficients at 250 Hz are

$$
\begin{array}{ll}
\text{glass} & 0 \cdot 1 \\
\text{wood} & 0 \cdot 15 \\
\text{plaster} & 0 \cdot 02
\end{array}
$$

$$\text{total absorption at windows} = 70 \times 0 \cdot 1\,\mathrm{m^2}$$

$$\text{floor} = 300 \times 0 \cdot 15\,\mathrm{m^2}$$

$$\text{walls and ceiling} = 330 \times 0 \cdot 02\,\mathrm{m^2}$$

$$\text{hence, } A = (7 \cdot 0 + 45 \cdot 0 + 6 \cdot 6)\,\mathrm{m^2}$$

$$= 58 \cdot 6\,\mathrm{m^2}$$

$$\text{Volume} = 30 \times 10 \times 5\,\mathrm{m^3}$$

$$\text{and } T = \frac{1500}{6 \times 58 \cdot 6}\,\mathrm{s}$$

$$= 4 \cdot 6\,\mathrm{s}$$

This would not be acceptable for public speaking.

Example 34

A sound of intensity $15\,\mathrm{W\,m^{-2}}$ at the surface of a partition wall is transmitted through it with attenuation 50 db. What is its intensity on the other side of the wall?

From the data

$$-50 = 10\lg\frac{I}{15}$$

and

$$I = 15\,\mathrm{antilog}\,(-5)\,\mathrm{W\,m^{-2}}$$

$$= 150\mu\mathrm{W\,m^{-2}}$$

Appendix D—Miscellaneous Examples for Practice in SI Units

EXAMPLES have been added to each section of the text. They are here presented in a random order, in the hope that students will attempt examples from all sections and hence appreciate the advantages of a common language of measurement which can sometimes bridge the apparent gaps between different disciplines.

Where substitution in a standard formula is required, the formula is usually given. Most other examples either repeat the worked examples or refer directly to points made in the text, so that acquaintance with the subject matter is not called for except at the most elementary levels. For the teacher who cares to select and work examples with a class, a few discussion points are added where appropriate. These questions have nothing to do with SI units, but help to rescue the examples from routine arithmetic.

In these examples take $g = 9{\cdot}81\,\mathrm{m\,s^{-2}}$; radius of the earth $= 6370\,\mathrm{km}$. Do not take results to significant figures not justified by the data.

1. The flow of the River Thames over the weir at Teddington varies considerably according to season, but its mean flow may be taken as $94\,\mathrm{kl\,s^{-1}}$. How long would it take for $1\,\mathrm{km^3}$ of water to pass over the weir and hence flow under London Bridge?

> *Query:* The Thames at London Bridge is tidal. In what way does this affect the result obtained? Is this result what one would guess without reference to the data?

2. If the earth and the moon were replaced by point charges each of $1\,\mathrm{C}$, what would be the force between them? Take the distance between their centres as $400\,000\,\mathrm{km}$.

> *Query:* Could such charges result in the moon's falling into the earth?

3. A man of mass $72\,\mathrm{kg}$ runs up 50 stairs, each rising $210\,\mathrm{mm}$, in 10 seconds. What is the average power he develops? If the energy expended were converted without loss into electrical energy, how long would it light a 40 W bedside lamp?

4. The mean depth of the total sea cover of the earth is given as 3800 m. Sea water has a mean density of 1020 kg m^{-3}. If the ratio of land area to sea area is 7:18, estimate the total mass of the oceans.

5. A four-cylinder oversquare four-stroke engine has a bore of 81 mm and a stroke of 58 mm. At what rate is air being drawn into the intake at 2500 rev/min? Assume that the cylinder pressure at the end of the induction stroke is that of the atmosphere. What mass of air does the engine require for a 10 hour run? Take the density of air as 1290 g m^{-3}.

6. Calculate the weight of an astronaut of mass 80 kg at a point in an orbit 350 km above the earth's surface, taking the weight of a small object to be inversely proportional to the square of its distance from the centre of the earth.

   Query: Does the result conflict with the phenomenon of 'weightlessness' experienced by astronauts in orbit?

7. A power station draws 18 Ml/hr of cooling water at 8°C from a river and discharges it at 15°C. What is the energy lost in one year?

8. The Great Pyramid, before it was stripped of its outer casing by the builders of Cairo, had a perpendicular height of 147·4 m and stood on a square base of side 251·6 m. Estimate its original mass if the limestone of which it is built has a density of 2500 kg m^{-3}.

9. A supersonic interferometer determines the wavelength of sound generated at very high frequencies in a small sample of liquid or gas. In liquid oxygen a frequency of 450 kHz gave a wavelength of 2432 μm. Given that the speed of a moving wave is the product of the frequency and the wavelength, what is the speed of sound in liquid oxygen?

10. The vertical force required to separate a flat horizontal plate of area A and mass M from the surface of water is given approximately by

$$F = Mg + 2A\sqrt{2gs}$$

where s, the surface tension is 72 mN m^{-1}. Find F for a flat disc of radius 100 mm and mass 50 g.

11. A mass of 1·4 kg hung on the end of a helical spring extends it by 32 mm. What is the work done in extending the spring by 75 mm, if it obeys Hooke's law that the extension is proportional to the extending force?

12. To start a car 200 A are drawn from the battery for 2 second. How many electrons then flow through the starter circuit?

13. Coils of fence wire are sold by mass. How many kilogramme are needed for a 3 strand wire fence to surround a field of perimeter 750 m. The diameter of the wire is 2·5 mm and its density is 7·7 Mg m^{-3}.

14. A car of mass 820 kg, using its engine as a brake, runs at a steady speed of 45 km/h down a hill of slope 1:10 (measured along the slope). What is the retarding force exerted on the car and at what rate is energy dissipated.

Query: In what form is the energy mainly dissipated, and where?

15. Two resistors of 3·4 Ω and 1·5 Ω are connected in parallel and the pair are then put in series with a resistor of 55 Ω and a 24 V battery of negligible internal resistance. Calculate the current flowing and the power dissipated in each resistor, and the total quantity of charge drawn from the battery after one hour.

16. A sluice gate for a flood control system has a mass of 5 Mg and resists a maximum total hydrostatic thrust of 1·2 MN. If the frictional forces in the guides amount to 10% of the total thrust, what is the minimum power that must be provided to raise the gate 1·3 m at a constant rate in 30 s. Take other mechanical losses as 50% of the indicated power.

Query: This question ignores any drop in the hydrostatic pressure as the gate is raised. Is this justified?

17. A 100 W lamp providing a total flux of 1270 lumen is fitted in an opal glass globe of 60% light transmission. What is the mean spherical candlepower of the fitting.

Note: MSCP is the luminous intensity calculated on the assumption that it is the same in all directions.

18. If water is drawn at the rate of 250 litre/min from a pipe of internal diameter 40 mm, what is the mean linear velocity of the water in the pipe?

19. The tunnels for the new Victoria Line of the London Underground have an outside bore of approximately 4·5 m diameter. What mass of spoil is removed in digging one kilometre of tunnel if the density of the clay stratum through which it is cut is 2400 kg m^{-3}. How many loads have to be run for each kilometre if each truck holds 4 m^3?

20. A non-magnetic toroidal former of mean diameter 300 mm is wound with 2500 turns of wire. What is the magnetic field strength produced in the torus by a current of 2 A flowing in the wire?

21. A 3 kW immersion heater fitted to a 100 litre domestic hot-water cylinder raises its temperature from 15°C to 85°C in 3 hours. What is the overall efficiency of the system?

22. A wooden beam of rectangular section 110 mm by 200 mm has a density of $920\,kg\,m^{-3}$. If it is cantilevered horizontally from one end, what is the shear stress at a point $5\cdot3\,m$ from the free end?

Note: the shear stress is the force per unit area acting parallel to the area.

23. Water is expelled from a horizontal pressure nozzle of diameter 20 mm at a velocity of $10\,m\,s^{-1}$, striking a fixed vertical plate. Obtain a first approximation to the pressure on the plate.

    Query: What assumptions are made to obtain this result and to what extent are they justified?

24. The osmotic pressure of a non-electrolyte in solution is directly proportional to the absolute temperature and the concentration. Hence if 1 mol of a substance is present in V litres of solution, the osmotic pressure is given by $P = RT/V$ where R is a constant. What is the osmotic pressure of a 5% solution of sucrose (M.W. $= 342$) at 15°C if R has the value $8\cdot3\,J\,mol^{-1}\,K^{-1}$.

    Query: The value given for R is identical with that for the Gas Constant. Why is this? Is the result dimensionally correct?

25. The run-on and run-off tensions of a belt driving a pulley of diameter $0\cdot75\,m$ at 950 rev/min are 20 N and 250 N. What power is being transmitted to the driven shaft?

26. Using the energy equivalents given in Appendix B, calculate how long 1 gramme of matter could run a 100 W lamp if its mass could be converted into electrical energy without loss.

27. A closed glass laboratory cage used for incubating insect eggs is 300 mm high and has an insulated base 300 mm × 600 mm. It is heated by an internal 40 W lamp. What will be the approximate air temperature inside the cage when it is 10°C outside? Take the heat loss through the top and sides as $5\,W\,m^{-2}\,K^{-1}$.

28. The first Whittle jet engine to be used successfully in flight had a fuel/thrust rating of $0\cdot14\,kg$ per newton hour. In a static thrust test lasting 5 hours the thrust was $3\cdot9\,kN$. How many litres of paraffin were consumed if its relative density is $0\cdot8$?

29. If the units as given in Appendix B for charge, energy, velocity, and action, i.e. e, (eV), c, $\left(\dfrac{h}{2\pi}\right)$ are taken as fundamental, is it possible to obtain expressions for derived units of mass, length, time, and current?

30. Two metal discs of diameter 0·75m are separated by an air gap of 3mm. What is the capacity of this system?

31. Find the mass of a small shot of radius 0·5mm and relative density 11·3. What vertical forces act on it if it is at rest in a fluid of relative density 1·2?

Stoke's Law gives, for the resistance at terminal velocity v of a small sphere of radius r falling through a viscous liquid of viscosity η

$$P = 6\pi\eta rv$$

What is the viscosity of a liquid of relative density 1·2 in which the terminal velocity of the shot is $12\,\mathrm{mm\,s^{-1}}$?

32. Although the value of the gravitational force at a point above the surface of a uniform sphere varies inversely as the square of the distance from its centre, at a point inside the sphere it varies directly as this distance. Calculate a value for g in a satellite orbit of radius 7500km, and at a point 2000km from the earth's centre.

> *Query:* Does a miner weigh less at the bottom of a mine shaft? Does the answer to the second part of Q.32 depend on a possible density distribution in the core and surface strata of the earth?

33. Calculate the work done in pumping dry a flooded cylindrical vertical shaft of diameter 5m and depth 50m. After the shaft is dry water begins to seep in at a rate of 20 litre/min. What is the power of the pump engine required to keep the shaft dry, if its overall efficiency is 20%?

34. Calculate the electric flux density at a point 1mm from an isolated electron.

> *Query:* In investigating atomic or nuclear forces, is it possible to ignore the presence of an electron at such a distance?

35. An elastic cord of length 2m is stretched to 4m by a force of 50N. How long would it be if stretched by the weight of a mass of 2kg?

36. An alloy-steel pressure pipe is 3m long and of internal and external diameters 57mm and 72mm. Find its mass if its material has a density of $7800\,\mathrm{kg\,m^{-3}}$.

37. A 5kW lighting set is installed in a farmhouse and driven by a petrol engine. What is its fuel consumption on 50% load if the overall efficiency of the installation is 10%? The relative density of the grade of petrol used is 0·71 and its heat of combustion $41\,\mathrm{MJ\,kg^{-1}}$.

38. A pile driver allows a mass of 250kg to fall 2m on a pile of mass 750kg. During the first few strokes it drives the pile 180mm into the ground at each stroke. This distance drops to 35 mm towards the end of the operation. Calculate the average resistance of the ground at the beginning and end of the pile driving, and the loss of kinetic energy at each impact.

39. An engine whose maximum power is 20 kW raises a vertical load of mass 500 kg. If the load starts from rest, for how long can it be accelerated at a steady rate of $1\,\mathrm{m\,s^{-2}}$?

40. The total output of exhaled water (as vapour) for an average man is 0·5 litre/day. The total area of the lungs and associated passages is about $70\,\mathrm{m^2}$. What is the average rate of evaporation in $\mathrm{mg\,s^{-1}\,m^{-2}}$?

41. The kinetic energy of a molecule of gas is proportional to its absolute temperature. If the mean kinetic energy of the molecules of a gas at 15°C is 0·03 eV per molecule, to what temperature must the gas be raised so that the mean molecule has an energy of 1 eV?

42. A glass marble of mass 10 g falls 350 mm on to a glass slab and bounces vertically with a coefficient of restitution of 0·95. What is the loss of kinetic energy at the impact?

43. An industrial diesel engine has a maximum power rating of 71·5 kW at 2400 rev/min. Its maximum torque is 240 N m at 1800 rev/min. What percentage of its maximum power is being developed at maximum torque?

44. Avogadro's hypothesis leads to the conclusion that 1 mol of any compound contains $6\cdot02 \times 10^{23}$ molecules. Taking graphite, of relative density 2·3, as monatomic carbon of atomic mass 12, find the volume of 1 mol. Hence find the volume occupied by one carbon atom in the graphite lattice.

> *Query:* If this calculation is repeated for other elements, do the volumes thus found bear any relation to the atomic masses?

45. A starter motor for a car engine has a running torque of 7 N m at 1000 rev/min. It then draws 270 A from its 12 V battery. What is its mechanical efficiency?

46. The cooling system of an internal combustion engine whose optimum running temperature is 91°C has a capacity of 3·4 litre on the engine side of the radiator thermostat valve. If the engine is started from cold at 0°C, what energy is absorbed by the water in reaching operating temperature.

47. The American Fall at Niagara is 50 m high. If all the energy released as the water falls appears as heat, what is the temperature difference at the top and bottom of the falls? Take the specific heat capacity of water as $4190 \, \mathrm{J \, kg^{-1} \, K^{-1}}$.

> *Query:* What are the other possible forms that potential energy of the water at the head of the fall is likely to take on falling? Would they be negligible in comparison with the heating effect?

48. A transistor mounted on a heat sink which provides a thermal coupling of $15 \, \mathrm{mW \, K^{-1}}$ to the surrounding air is used to control a constant $500 \, \mathrm{mA}$ $12 \, \mathrm{V}$ supply from a $15 \, \mathrm{V}$ input. If the maximum temperature rating of the transistor is $180°\mathrm{C}$, what is the maximum permissible ambient air temperature?

49. According to the Heisenberg principle the minimum uncertainty in the measure of energy over a small time δT is given by

$$\delta E \times \delta T = h/2\pi = 10^{-34} \, \mathrm{J \, s}.$$

What is the value of δE (in electron volt) for an atomic transition of $10^{-8} \mathrm{s}$?

50. The mass of a proton is $1 \cdot 67 \times 10^{-27} \, \mathrm{kg}$. Using the Einstein expression for the total energy equivalent $E = mc^2$, calculate the energy required to create a proton.

51. A $1000 \, \mathrm{MW}$ power station would use coal fuel at a rate of $10 \, 000$ Mg/day. What is its overall efficiency if coal has a specific energy of $28 \cdot 3 \, \mathrm{MJ \, kg^{-1}}$?

52. An inland power station using evaporative cooling towers would evaporate about $3000 \, \mathrm{l/MW}$ of water each hour. If the latent heat of water vapour over the temperature range concerned is approximately $22 \, \mathrm{J \, kg^{-1}}$, how much energy is thus dissipated?

Bibliography

FURTHER information on the definitions of the units, the history of the proposals, and equivalent values in non-SI units may be obtained from:
Anderton and Bigg: *Changing to the Metric System*. HMSO (1967).
The following publications of the British Standards Institution discuss formal definitions and agreed usage.

BS 3763:	1964	*The International System (SI) Units.*
PD 5686:	1965	*The Use of SI Units.*
BS 1637:	1950	*The M K S System of Electrical and Magnetic Units.*
BS 2990:	1958	*Rationalized and Unrationalized formulae in Electrical Engineering.*
BS 1991:		*Letter symbols, signs and abbreviations.*
	Part 1	*General.*
	1967	
	Part 6	*Electrical Science and Engineering.*
	1963	

The following discusses usage at school levels
Association for Science Education: *SI Units, Signs, Symbols and Abbreviations*. Cambridge, 1969.
and is based on The Royal Society: *Symbols, Signs and Abbreviations*. London 1969.

Answers to Miscellaneous Examples

RESULTS are given to 1, 2 or 3 significant figures according to the data of the problem.

1. 120 days
2. $5 \cdot 6 \times 10^{-8}$ N
3. 742 W; 3 min
4. $1 \cdot 4 \times 10^{21}$ kg
5. $0 \cdot 025$ m^{3}s^{-1}; 1160 kg
6. 705 N
7. $4 \cdot 5 \times 10^{15}$ J
8. $7 \cdot 8 \times 10^9$ kg
9. 1090 ms^{-1}
10. $0 \cdot 57$ N
11. $1 \cdot 22$ J
12. $2 \cdot 5 \times 10^{20}$
13. 85 kg
14. 805 N; 10 kW
15. 55 Ω, 430 mA, 10 W; $3 \cdot 4$ Ω, 130 mA, $57 \cdot 5$ mW; $1 \cdot 5$ Ω, 300 mA, 135 mW; 1540 C
16. $14 \cdot 6$ kW indicated
17. $60 \cdot 7$ cd
18. $3 \cdot 3$ ms^{-1}
19. 38 000 Mg; 4000 loads
20. 5310 Am^{-1}
21. 90%
22. $47 \cdot 9$ kN m^{-2}
23. 31 N
24. 349 Nm^{-2}
25. 8600 W
26. $2 \cdot 8 \times 10^4$ years
27. 21°C
28. 3400 l

29. mass: (eV) c^{-2}; length: c (eV) $\left(\dfrac{h}{2\pi}\right)^{-1}$; time: $\left(\dfrac{h}{2\pi}\right)$ (eV)$^{-1}$; current: e (eV) $\left(\dfrac{h}{2\pi}\right)^{-1}$
30. $1 \cdot 3$ nF
31. $5 \cdot 9$ mg; gravity 58 μN, hydrostatic lift $6 \cdot 2$ μN; $0 \cdot 46$ Nsm^{-2}
32. $7 \cdot 02$ ms^{-2}; $3 \cdot 05$ ms^{-2}
33. at least 241 MJ; 820 W
34. $1 \cdot 27 \times 10^{-14}$ Cm^{-2}
35. $2 \cdot 79$ m
36. $35 \cdot 5$ kg
37. $3 \cdot 1$ l/hr
38. $16 \cdot 4$ kN; $43 \cdot 8$ kN; $3 \cdot 3$ kJ
39. $4 \cdot 1$ s
40. $0 \cdot 083$ mg s^{-1} m^{-1}
41. 9330°C
42. $3 \cdot 35$ mJ
43. $63 \cdot 3$%
44. $8 \cdot 64 \times 10^{-30}$ m^3
45. $22 \cdot 6$%
46. $1 \cdot 26$ MJ
47. $0 \cdot 12$°C
48. 80°C
49. $6 \cdot 25 \times 10^{-6}$ eV
50. 940 MeV
51. $30 \cdot 5$%
52. 66 kJ/MWh

Index

acceleration, 5, 35
action, 38
adiabatic changes, 19
ampere, 23
   turn, 29
angle, 6
angular acceleration, 6, 37
angular momentum, 38
angular velocity, 6, 35
area, 7
atmosphere, standard, 11

bar, 11

calorific value, 14
candela, 20
capacitance, electrostatic, 28, 36
capacity, fluid, 7
Celsius, 16
centigrade, 6, 16
charge, 23, 36
coulomb, 23, 36
current, 22

decibel, 40
decimal point, 5, 7, 14
degree, Celsius, 16
   Kelvin, 16
   sexagesimal, 6
density, 9, 35
dialectric constant, 27
dimensions, 34

electric field strength, 26
electric flux density, 25
electrolysis, 24
electron, 37
   volt, 38
energy, 9, 36
   bond, 14

energy, field, 24
   heat, 14
   specific, 14
entropy, 19, 35

farad, 28, 36
force, 10, 36
   electromotive, 23, 36
   electrostatic, 28
   magnetomotive, 29
flux, electric, 25
   luminous, 20, 36
   magnetic, 30, 36

'g', 10
gas constant, 16
Gaussian units, 32
Gauss theorem, 20, 25
Giorgi, 22
grade, 6
gramme, 9
gramme molecule, 15
gravitation, 10

heat, 14, 36
   conduction, 18
   latent, 17
   loss, 18
   specific, 17
Heaviside–Lorentz units, 32
hectare, 7
henry, 31, 36
hertz, 8
horse-power, 1, 34

illumination, 21, 36
impulse, 9
inductance, 31, 36
inertia, 9
intensity, light, 20
   sound, 39

joule, 11, 36

Kelvin, 16
kilogramme, 9

light year, 4
litre, 7
lumen, 20, 36
lux, 21, 36

magnetic field strength, 30
magnetic flux density, 30, 36
magnetic force, 29
mass, 9
metre, 4
metre-kilogramme-second system, 22
micron, 4
mil, 7
millimetres of mercury, 11
modulus, Young's, 11
mole, 15
moment, 12
momentum, 9

newton, 10, 36
notation, index, 2
    solidus, 2, 17

ohm, 24, 36

permeability, 32, 35
permittivity, 27, 33, 35
Planck's constant, 37
potential difference, 26, 36
potential gradient, 26
power, 16, 36
prefixes, 2
pressure, 11, 35

quantum, 38

radian, 6
resistance, 24, 36
    specific, 24
reverberation, 39
rotation, 6

second, 4
solid angle, 7
spin, 38
steam, 17
steradian, 7
surface tension, 13, 35

temperature, 15
tesla, 30, 36
thermal conductance, 18
time, 4
ton, 9
tonne, 9
transformer, 24

U value, 18
unit symbols, 2
units, customary, 1
    derived, 1, 35, 36
    fundamental, 1
    multiple, 2
    nautical, 7
    submultiple, 2

velocity, 5, 35
viscosity, 12, 35
volt, 23, 36
volume, 7, 35

watt, 11, 23, 36
weber, 31, 36
weight, 10
work, 11, 36